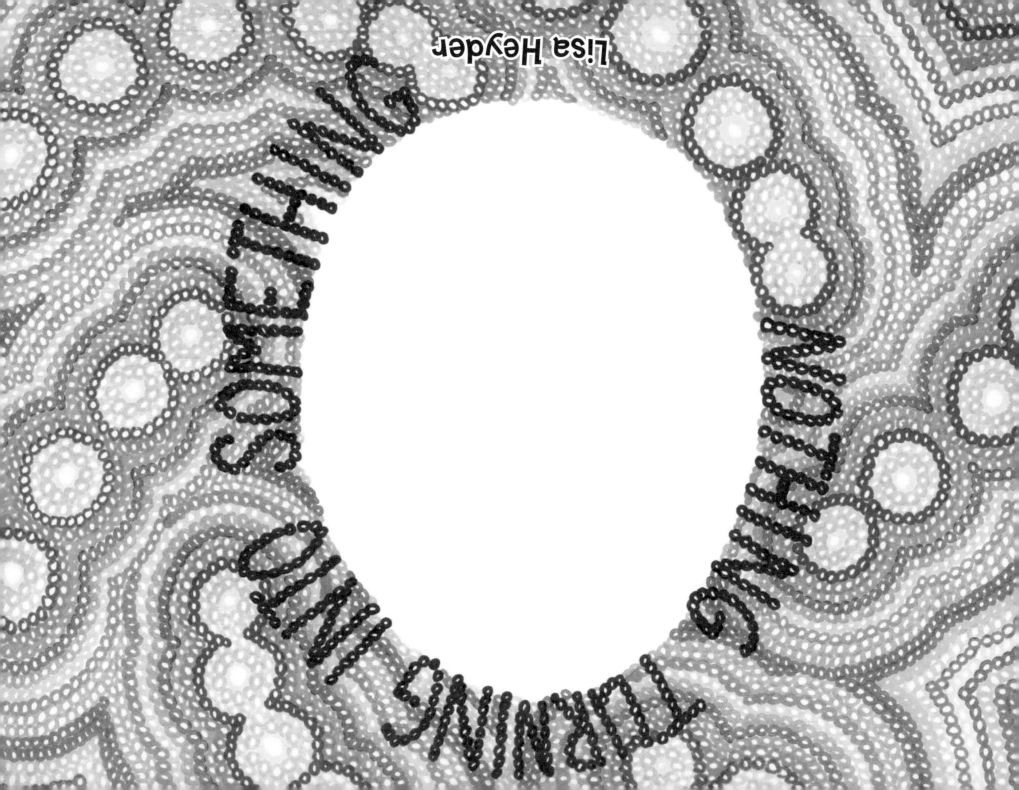

SOMETHING TURNING INTO SOMETHING

Lisa Heyder

 www.trafford.com

North America & international
toll-free: 1 888 232 4444 (USA & Canada)
fax: 812 355 4082

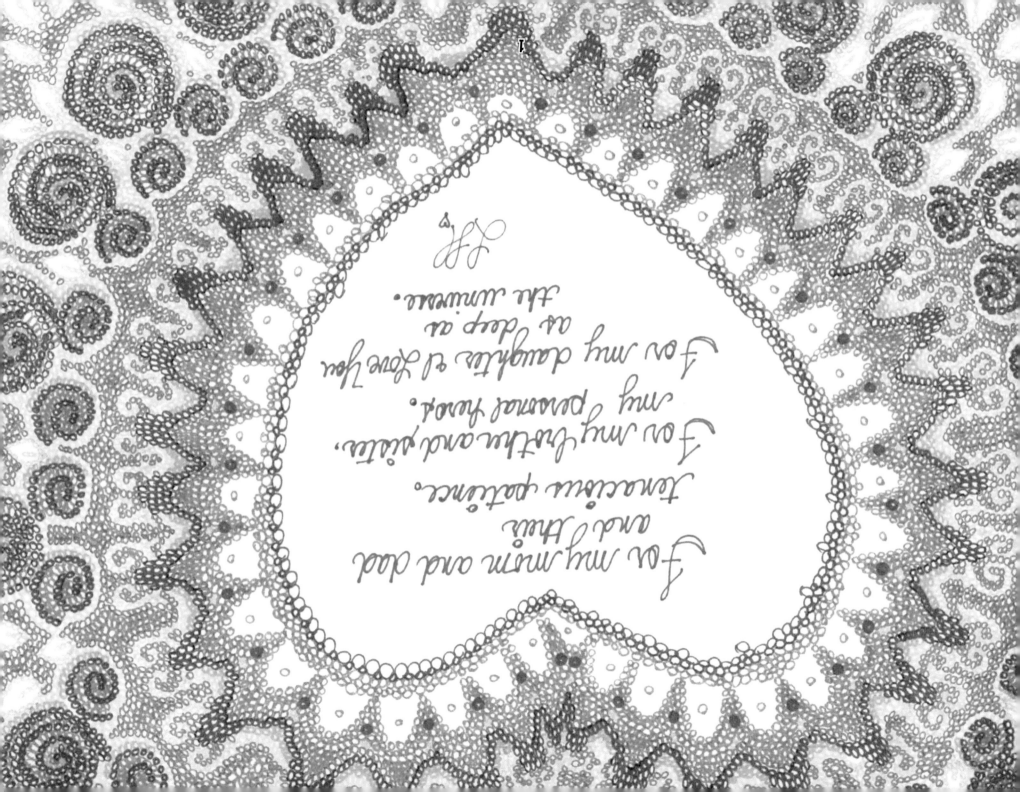

For my mum and dad
and their
tenacious patience.

For my father and sister,
my present hero.

For my daughter I love you
as deep as
the universe.

Nil

Zero

This is Nothing.

Diddly-squat

Nada

Zilch

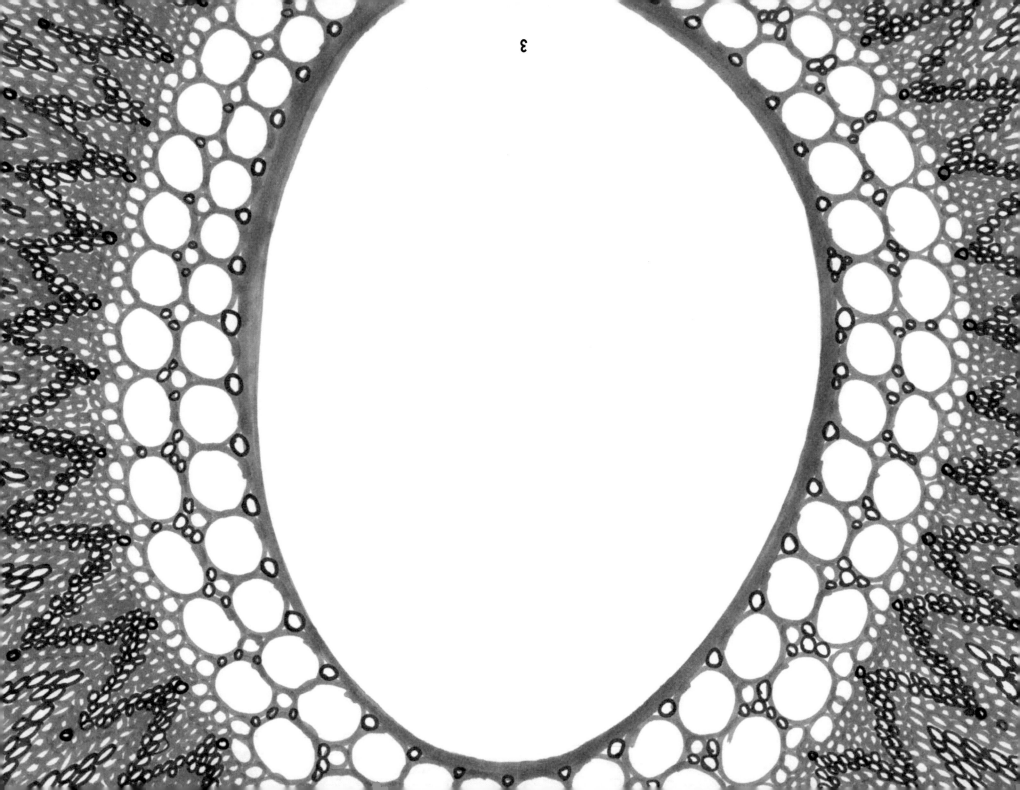

Feeling invisible, small, neglected, and alone, Nothing was sad and empty.

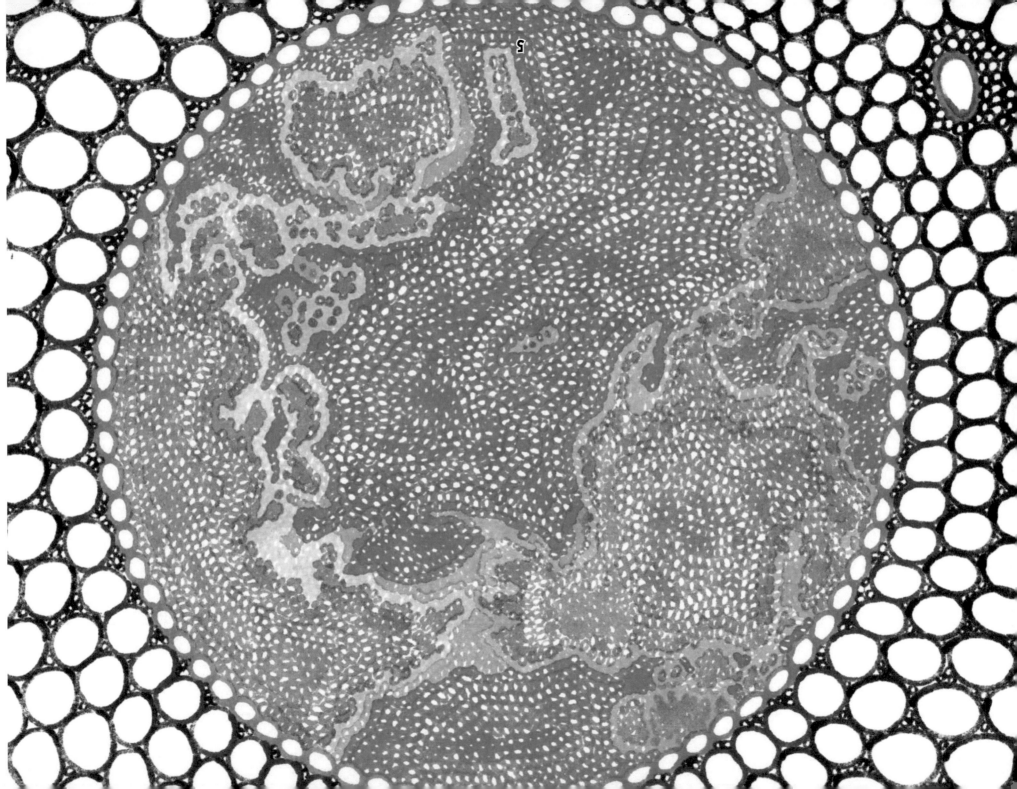

All the numbers were fuddle-duddling. Little did they know they were missing someone **important.**

fundamental
essential
vital
relevant
significant
valuable

Nothing started thinking.

pondering
ruminating
studying
considering
speculating
deliberating
figuring

Nothing had a notion
feeling
idea
belief
hunch

that the others possessed

something that Nothing

did not.

For a long time, Nothing sat still and reflected into Nothing.

As Nothing reached inward, Nothing saw into infinity.

In that moment, Nothing discovered the majestic being inside.

sacred
ubiquitous
all powerful

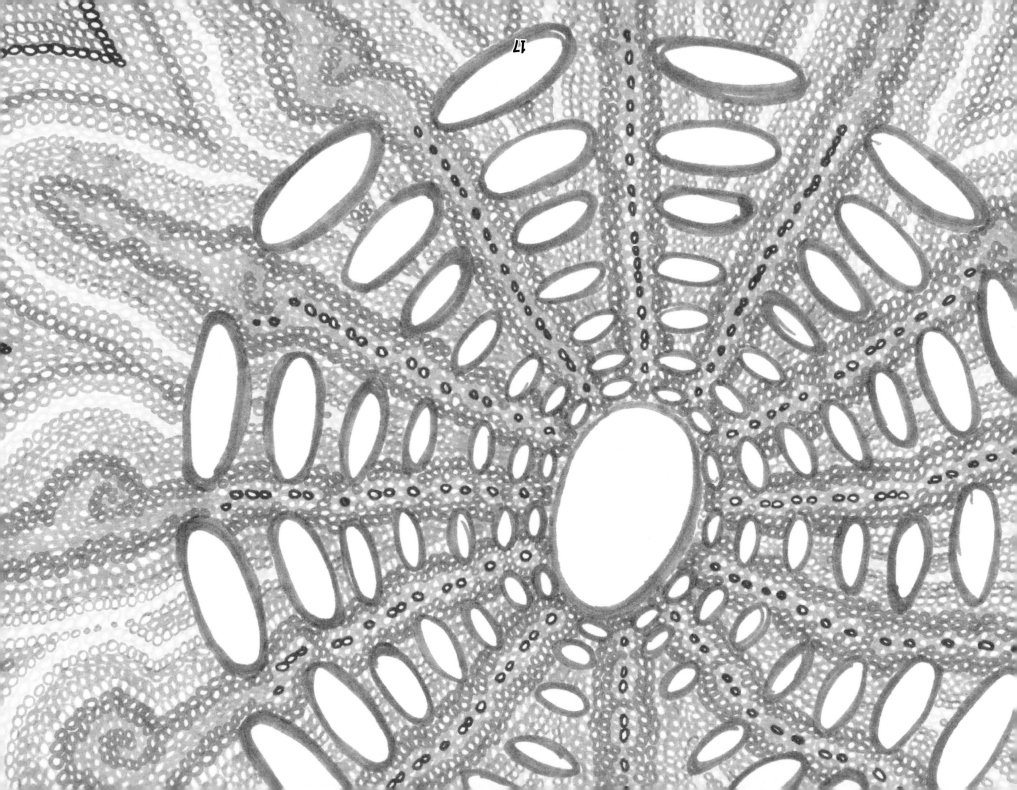

Nothing realized the others would not get very far minus Nothing.

subtracting

As the others got organized
Nothing decided to meet them
all.

greet
f
face
join
address

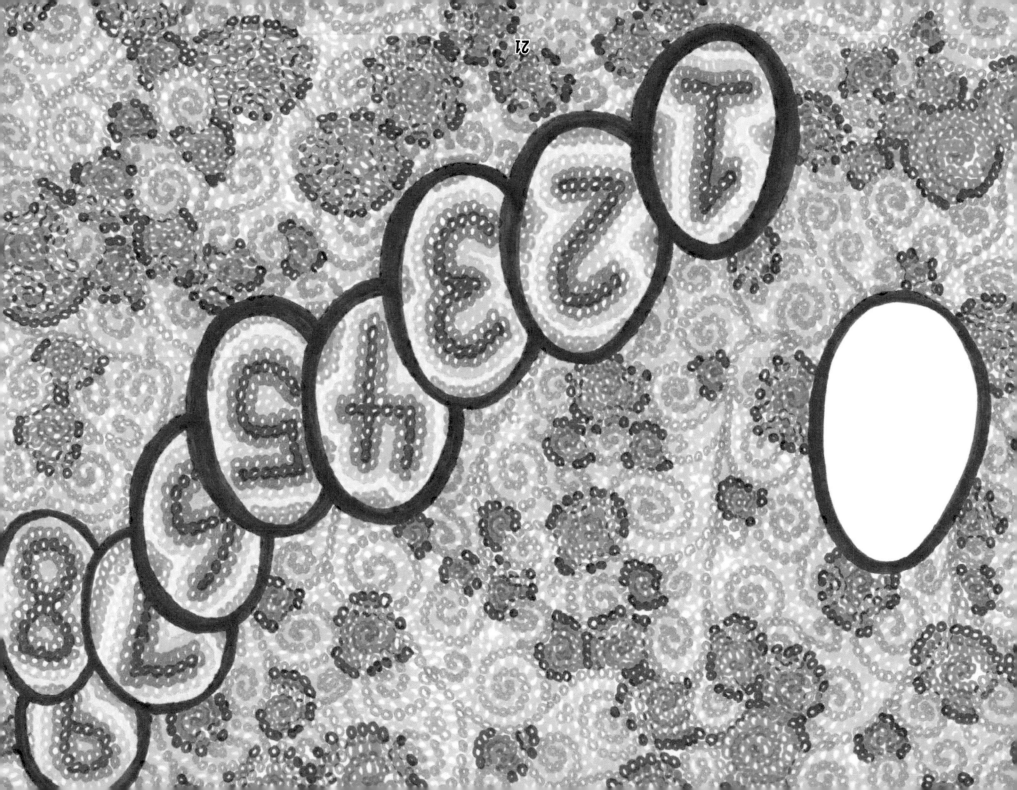

Nothing rolled over to One, and said, "I would like to meet everyone." Oddly, One understood but noticed Nothing in front of One did not change One.

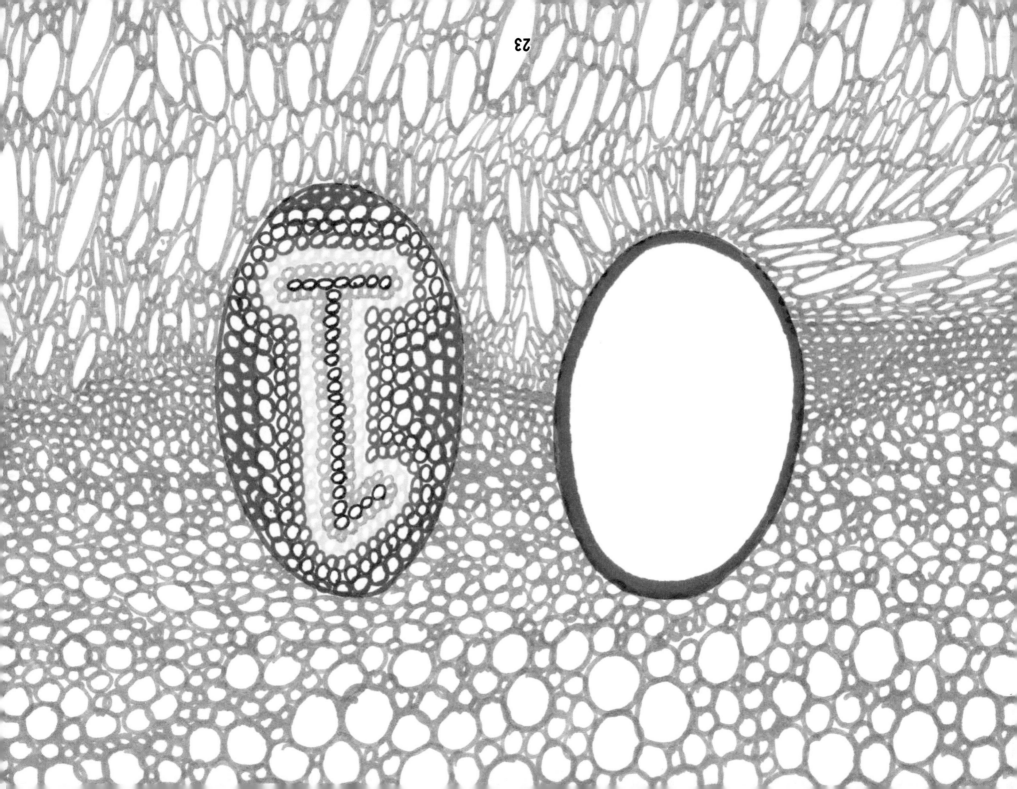

As Nothing continued meeting and greeting, even Two seemed easy to cooperate with.

work
play

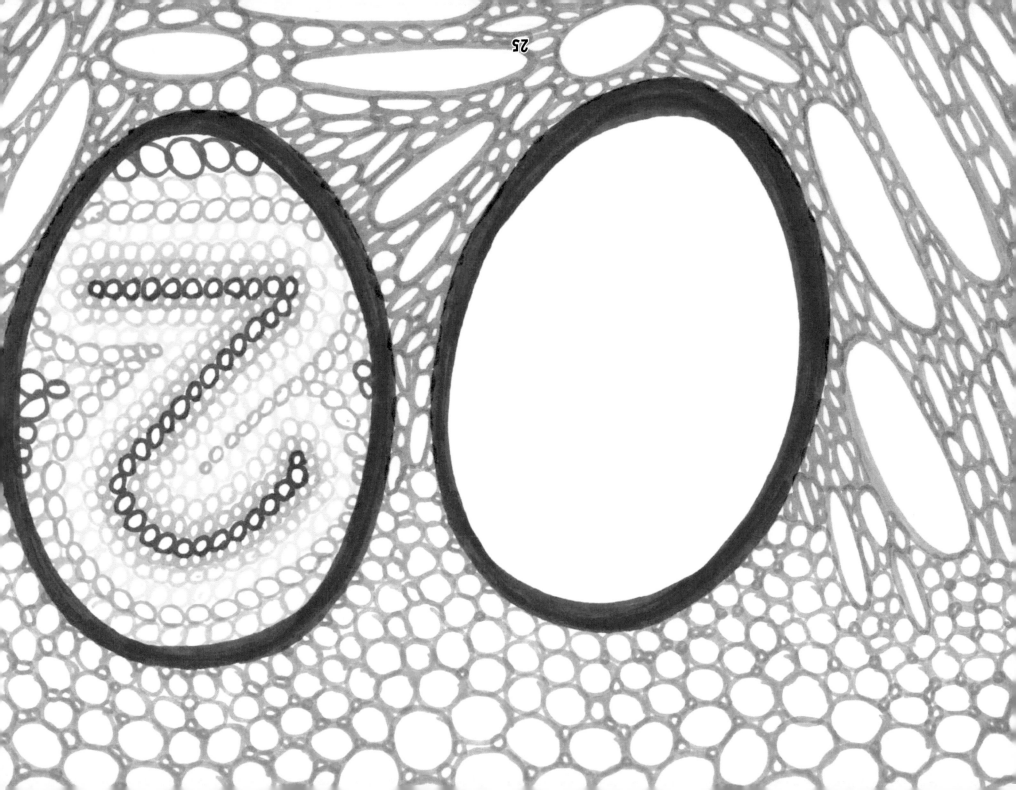

Oddly, Three seemed a little more ~~challenging~~ demanding. Even Four seemed difficult, but reminded Nothing of Two.

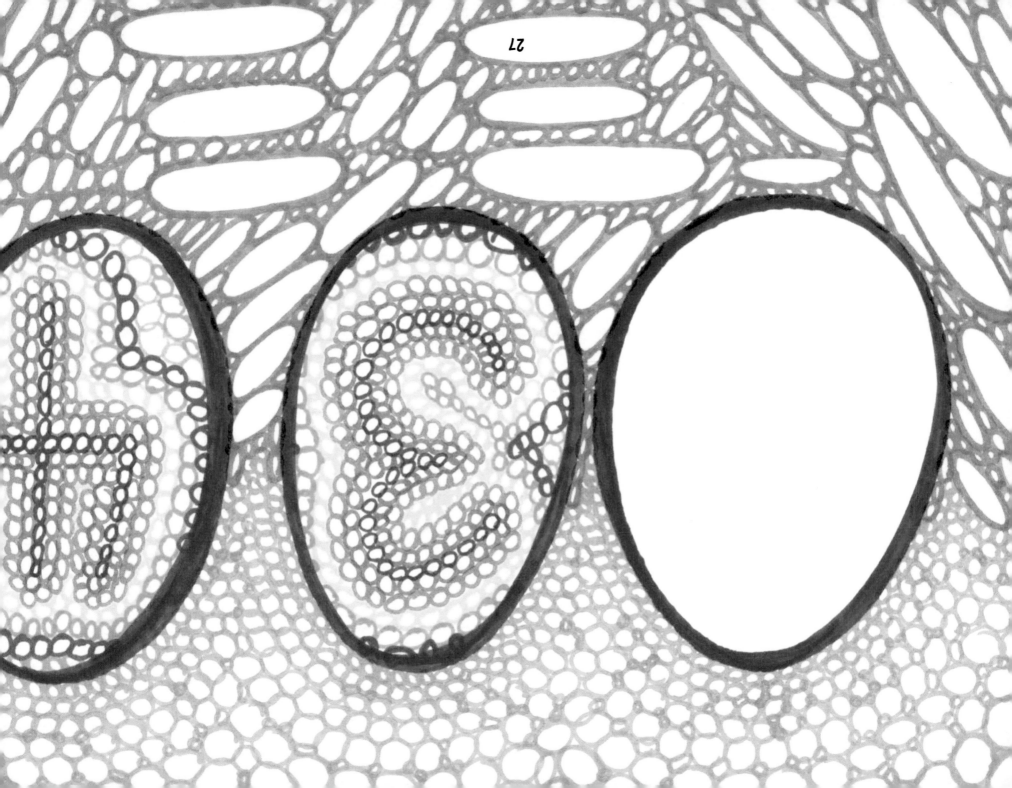

Oddly, Five was easy to play with. Somehow, even Six reminded Nothing of Two and Three. Seven oddly appeared to be a stick in the mud.

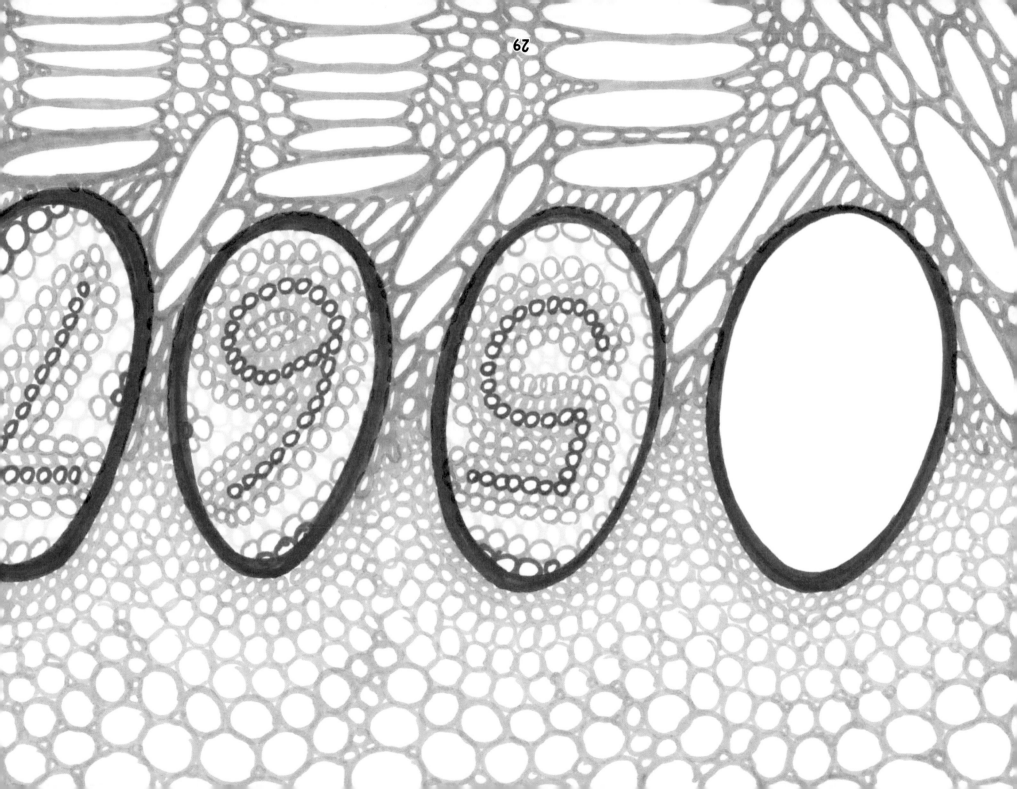

Even Eight reminded Nothing of Two. Oddly, Nine reminded Nothing of Three. Nothing was amazed by the similarities and differences between each number.

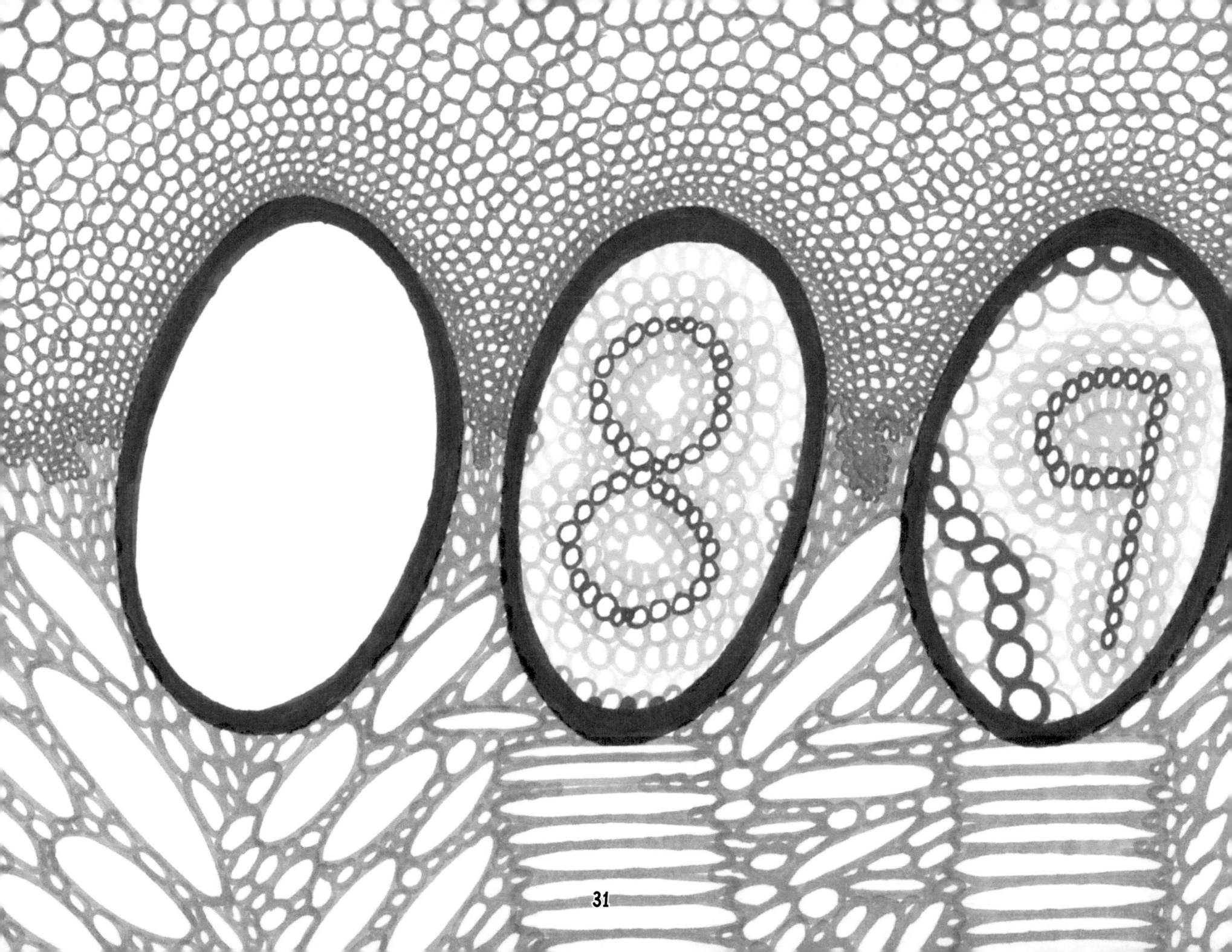

Nothing started to grow. One said, "You are such a well rounded member that every number disappears when multiplied by your roundness."

wholeness

The question became,
"What next...?" and Nothing
thought...

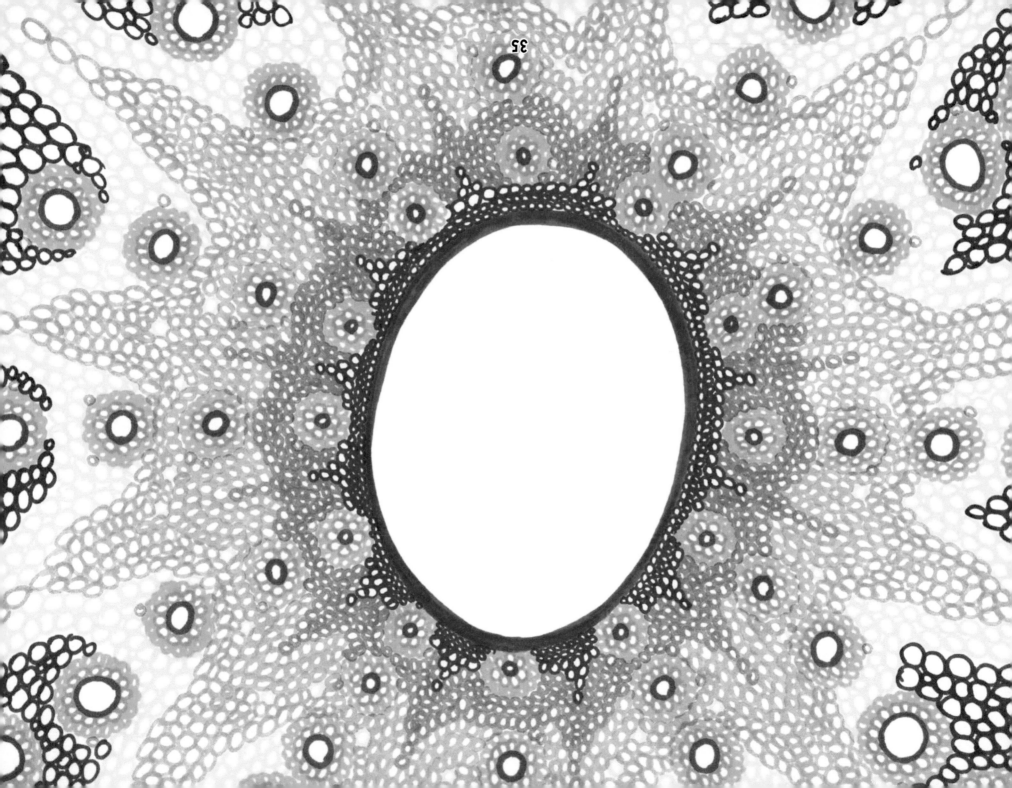

A joule of energy struck
Nothing, and Nothing exclaimed
declared
emitted
proclaimed
"One, you go first now!"

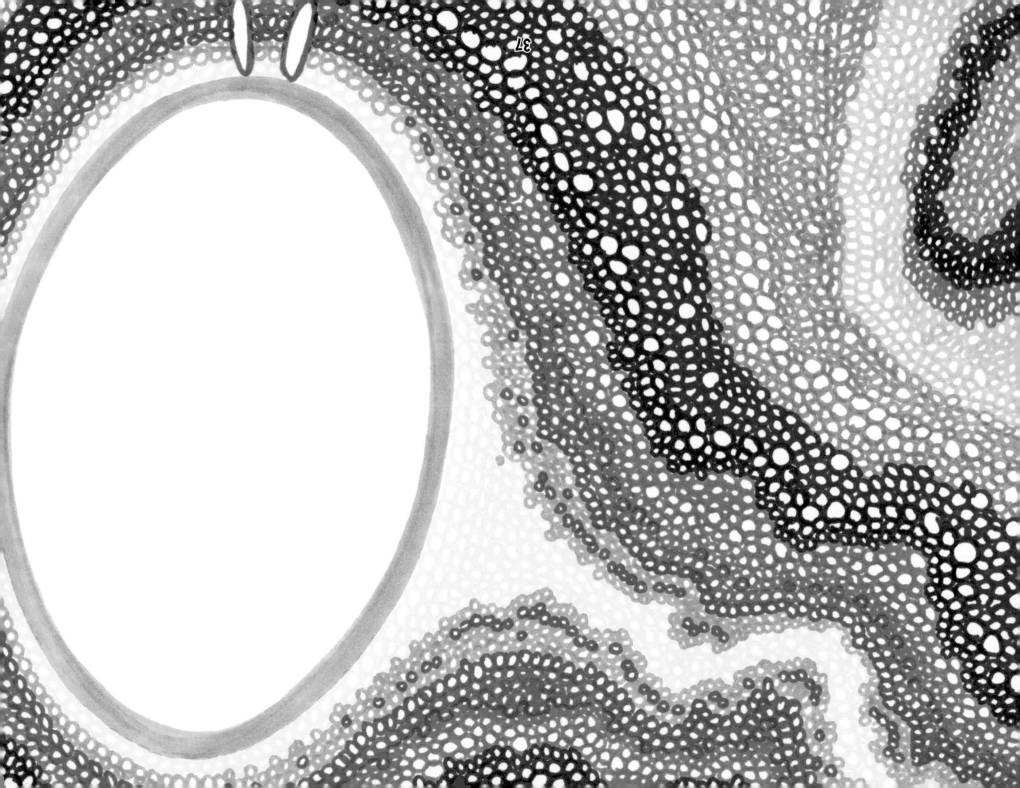

As One took the lead,
the joule of energy in Nothing
engulfed One as they came
together and ≥POW≤ they

became Ten.
turned into
metamorphosed into
changed into

Both One and Nothing felt stronger and more powerful than ever before.

Wait! Nothing noticed something! Nothing noticed there was a secret. Ahaa! Nothing was and is the

starting point

orgin.

beginning

When the others followed
Nothing, they could all
see into infinity.

Nothing had become a force, Nothing mattered and started accelerating toward Nothing's potential.

And, in this time and space, Nothing absolutely truly genuinely mattered. Nothing could relate to all.

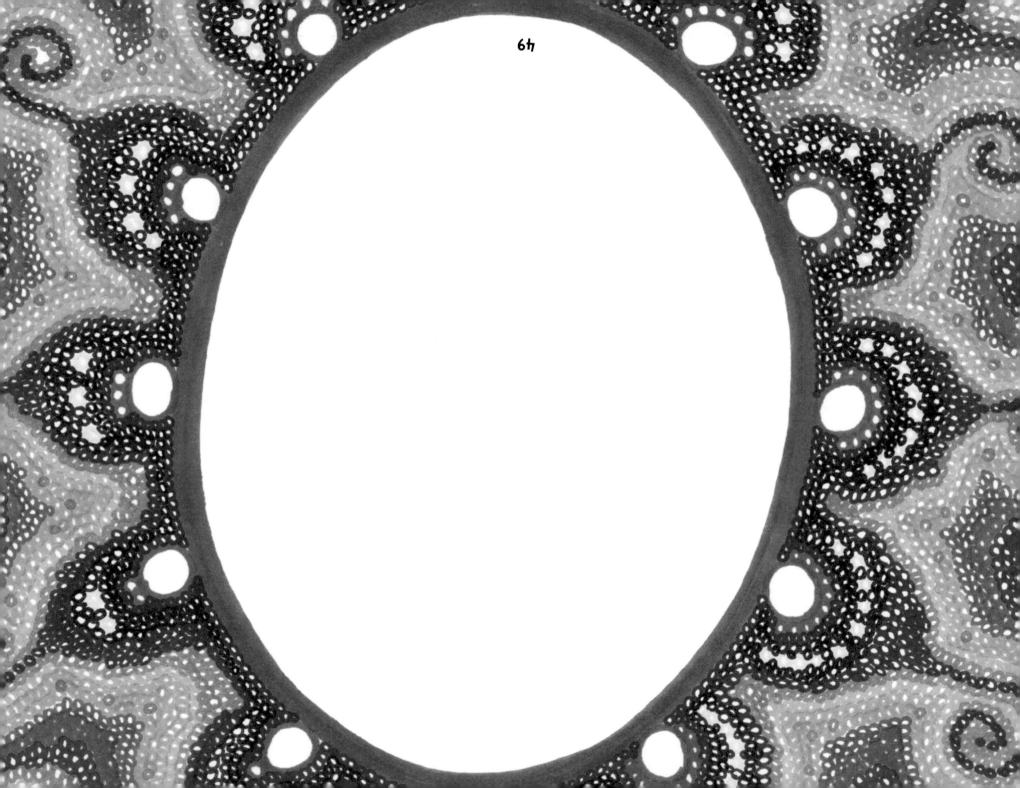

Nothing became Something!

Nothing was Something!

Nothing is Something!

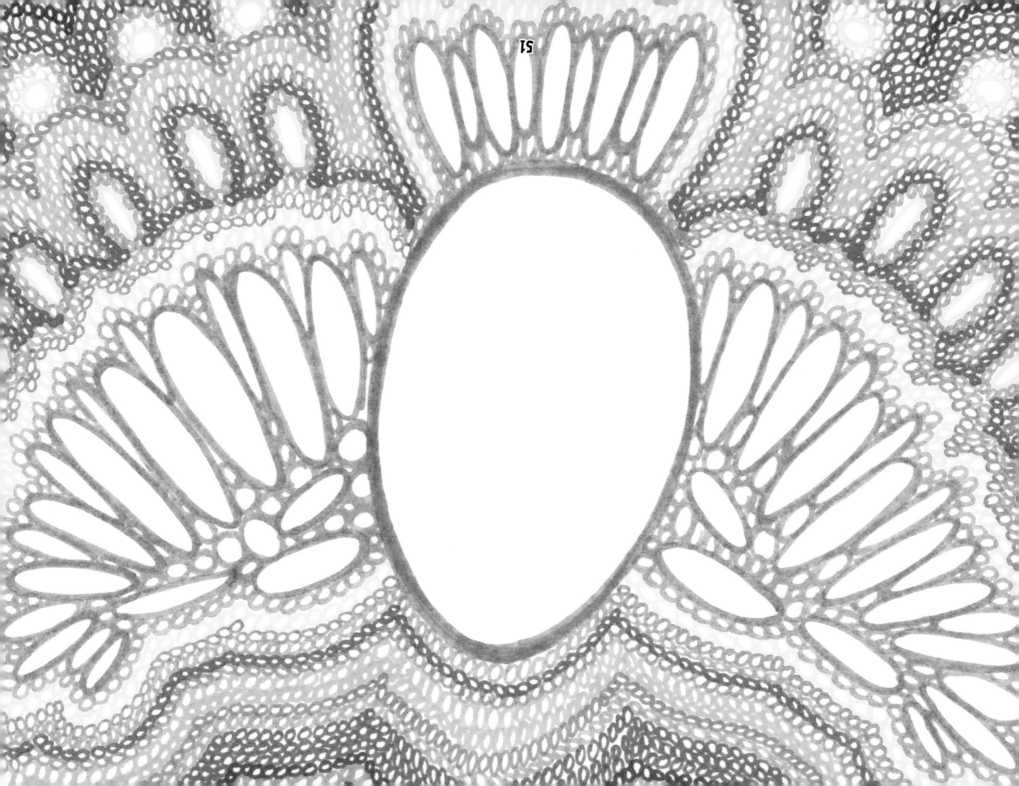